BEASTLY BATTLES

HIPPOPOTAMUS VS. STEGOSAURUS

Published by Capstone Press, an imprint of Capstone.
1710 Roe Crest Drive, North Mankato, Minnesota 56003
capstonepub.com

Library of Congress Cataloging-in-Publication Data
is available on the Library of Congress website.
ISBN: 9781669088721 (hardcover)
ISBN: 9781669088936 (paperback)
ISBN: 9781669088769 (ebook PDF)

Summary:
Don't get it their way! These two giants are ready to clash! The hippopotamus is one of the toughest beasts today. But millions of years ago, Stegosaurus had powerful defenses against even the most dangerous dinosaurs. Make sure to stand a safe distance away. These two mighty beasts are ready to collide in an epic battle!

Editorial Credits:
Editor: Peter Mavrikis; Designer: Bobbie Nuytten; Media Researcher: Svetlana Zhurkin; Production Specialist: Whitney Schaefer

Image Credits:
Capstone: Jon Hughes, 5 (bottom), 10, 14, 18, 19, 22, 27 (bottom); Dreamstime: Andrewmolinaro, 25, Markoflaherty, 21; Getty Images: Mark Newman, 9, Sciepro, 29, Stocktrek Images/Elena Duvernay, 23, Stocktrek Images/Mark Stevenson, 15, Winfried Wisniewski, 20; Shutterstock: AlexDibnah, 13, Bruce Ellis, 24, Chedko, 27 (top), Daniel Eskridge, 7, 11, Dmitriy Melnikov, 28, Fadil Youness, 17, Lena Ivanova, 8, manych, 16, MartinMaritz, 12, nwdph, 6, PhotocechCZ, cover (top), 5 (top), Warpaint, cover (bottom)

Printed and bound in China. PO 6098

TABLE OF CONTENTS

Words in **bold** are in the glossary.

BATTLE OF THE BIG BEASTS

It's a battle of the big bad beasts.

One lived long ago and had several powerful **defenses**.

The other is one of the fiercest animals living today. Which deadly giant will win this big battle?

HIPPOPOTAMUS

STEGOSAURUS

MEET THE COMBATANTS

The hippopotamus and Stegosaurus are **herbivores**. They only eat plants. But don't let that fool you. They are deadly beasts!

Both the hippopotamus and Stegosaurus are huge. Their size is both a weapon and a defense. Few **predators** would go after these big beasts.

RIVER HORSE

Hippopotamus means "river horse." "Hippos" are dangerous in the water. They are too heavy to swim, but watch out! A hippo can move really fast underwater.

Hippos have eyes and ears on top of their heads. This helps them stay underwater where they can keep watch for predators.

DANGEROUS DINO

Stegosaurus lived during the Late Jurassic Period. This was the age of dinosaurs. Stegosaurus was surrounded by dangerous dinos like the mighty Allosaurus.

Stegosaurus was huge. Scientists believe an adult weighed about 14,000 pounds (6,350 kilograms). That's the same weight as a school bus!

BIG AND BAD

The hippo is one of the largest land animals in the world. An adult hippo can weigh up to 10,000 pounds (4,500 kg).

No animal is going to mess with an adult hippo. Not even a hungry lion. However, baby hippos are **vulnerable**. They can be attacked by crocodiles or hyenas.

ARMORED ANIMAL

Stegosaurus was one tough beast. It had to protect itself from deadly dinos.

The skin of Stegosaurus was covered in hard **scales**. This giant also had large **plates** along its back. Even Allosaurus had trouble biting into Stegosaurus.

CHOMP!

The hippo's deadliest weapon is its teeth. They're huge! A hippo's teeth can be up to 20 inches (51 centimeters) long.

The hippo also has a powerful bite. Its bite is one of the strongest in the animal kingdom. A hippo can snap bones with one bite.

WHIP IT!

Stegosaurus had a long, powerful tail. The end of its tail was covered in sharp spikes. If a predator approached, it could whip its tail at the attacker. *Ouch!*

It's a good thing Stegosaurus had these defenses. This dino wasn't too smart. It had a tiny brain for such a big creature. Its brain was the size of a walnut!

PROTECTIVE PARENTS

The hippo only eats plants. However, it's still one of the deadliest animals in Africa!

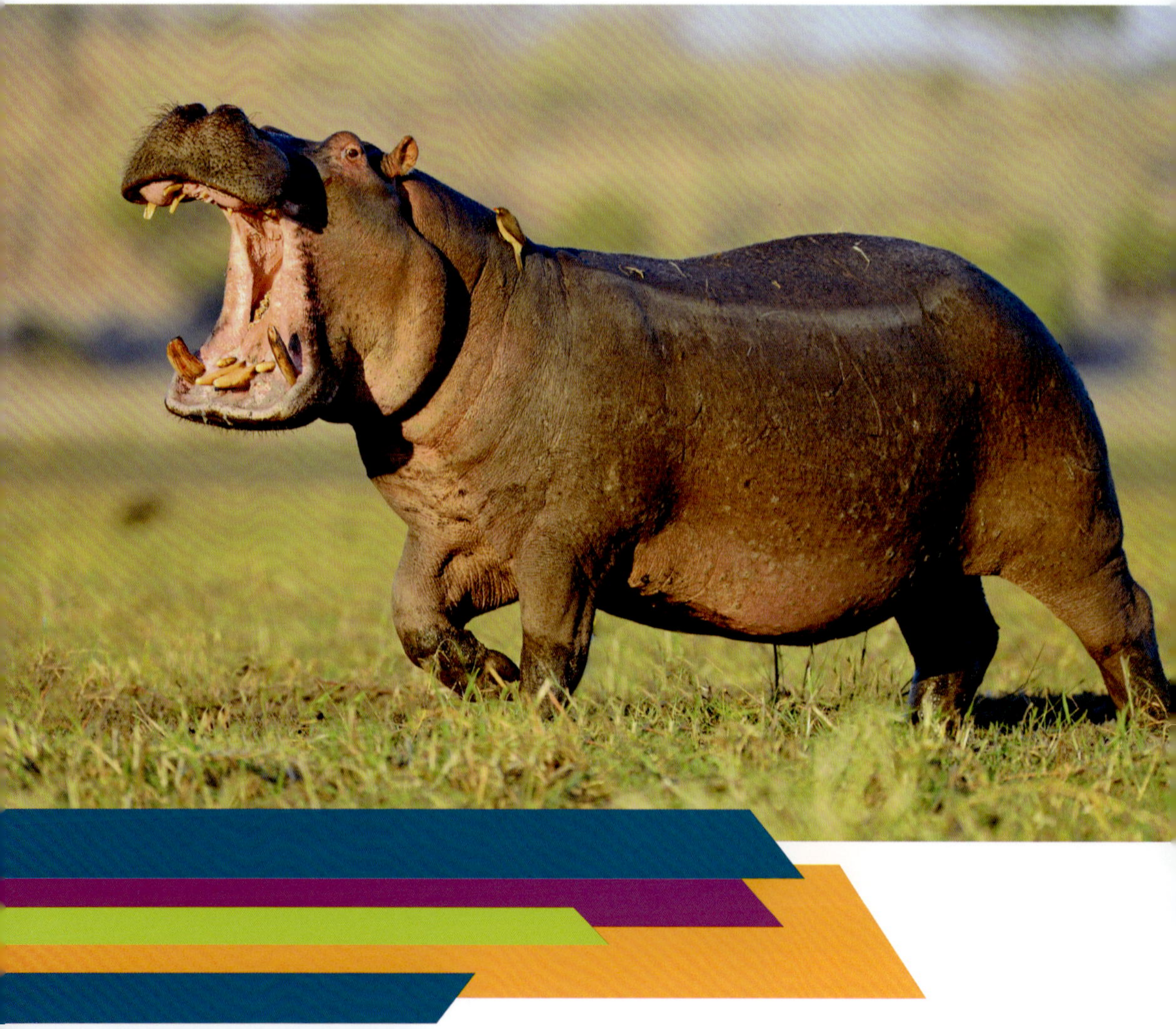

Hippos are deadly because they are **aggressive**. They will protect their young at all costs. Hippos are ready to attack anything that threatens them.

TAKING A STAND

Stegosaurus had very short legs for such a large animal. Because of this, the giant lizard was probably very slow.

Good thing Stegosaurus had so many defenses. This slow dino would need to stand its ground. Stegosaurus could never outrun a hungry predator!

SAFETY IN NUMBERS

The hippopotamus is a **herd** animal. Most herds have 10 to 30 hippos. Some herds can have as many as 200 hippos!

This **strategy** helps them scare away deadly predators. Not even a lion or leopard would attack a group of angry hippos.

FIGHT!

The hippo makes its way out of the wetlands. It opens its mouth wide to show off its dangerous teeth. Hippos could bite through Stegosaurus's tough scales.

But Stegosaurus isn't afraid. It stands its ground. Watch out for that tail! Those spikes are deadly.

STRONG TEETH AND JAWS

TAIL SPIKES

WHO'S THE WINNER?

You've met these two massive beasts. The hippo is a living giant with a bad attitude. Stegosaurus survived being hunted by the deadliest dinos on the planet. Who would win this battle between big beasts?

	Hippopotamus	Stegosaurus
RANGE	Africa	western North America
HABITAT	wetlands	wooded areas
HEIGHT	over 5 feet (1.5 meters)	up to 14 feet (4.3 m)
WEIGHT	up to 10,000 pounds (4,500 kg)	about 14,000 pounds (6,350 kg)
SPEED	• 22 miles per hour (35 kilometers per hour) on land • 5 miles per hour (8 kph) underwater	5 miles per hour (8 kph)
WEAPONS	• long curved teeth • powerful bite	• spiked tail • thick skin
STRATEGIES	• stays in the water • herd animal • aggressive	• stood its ground • swung tail for defense

GLOSSARY

aggressive (uh-GREH-siv)—being forceful or strong

defense (dih-FENSS)—a way to protect

herbivore (ERB-iv-or)—an animal that feeds only on plants

herd (HERD)—a large group of animals that lives together

plate (PLAYT)—a thin, flat material used for protection

predator (PRED-uh-tur)—an animal that eats other animals

scale (SCAYL)—a small, thin plate that covers skin

strategy (STRA-tuh-gee)—a plan

vulnerable (VUL-ner-ah-bul)—open to attack or injury

READ MORE

Anthony, William. *Hippopotamus*. New York: Enslow Publishing, 2022.

Riggs, Kate. *Hippopotamuses*. Mankato, MN: Creative Education/Creative Paperbacks, 2022.

Sabelko, Rebecca. *Stegosaurus*. Minneapolis: Bellwether Media, 2020.

INTERNET SITES

The Dinosaur Database
dinosaurpictures.org

National Geographic Kids: Hippopotamus
kids.nationalgeographic.com/animals/mammals/facts/hippopotamus

Natural History Museum: A Stegosaurus Brought to Life
nhm.ac.uk/discover/dino-directory/stegosaurus.html

INDEX

ABOUT THE AUTHOR

Charles C. Hofer is a writer and biologist living in New Mexico, not far from where Stegosaurus once roamed (but very far away from any hippos).